I0488168

Kids Count by Groups

By Nancy Radke

Illustrated by

Eden Radke & Alison Ralston

ISBN 13 978-1500965662
ISBN 10 1500965669

Bedrock Distribution LLC
9805 NE 116th Street, PMB #7362
Kirkland, WA 98034 USA
Printed in USA

Why should I learn to count by groups?

When adding and subtracting, we work with single items. In other words, we count by ones.

When we multiply and divide, we work with groups of items. It is easy to multiply with tens, twos, or fives, as we've often counted by those numbers.

We count frog eyes by twos.
2, 4, 6, 8, 10.

Here's a chance for a child to count by groups of threes, fours, sevens, and nines, as well as the other numbers up to twelve.

Backgrounds are colored the same color to help memory.

Make sure your child counts in the right order, so the answers match.

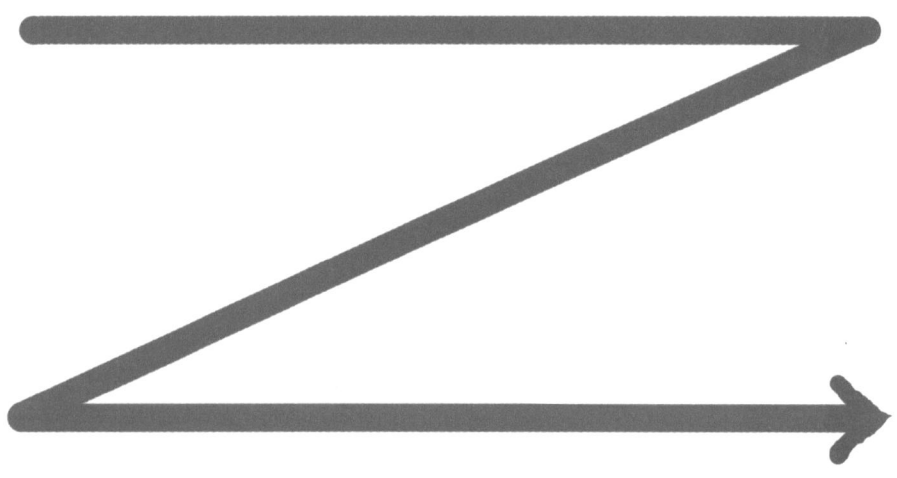

Follow the arrow. Count by ones to five.

Always count in the same direction, so you don't
lose your place.

Count by ones to ten.

Count by ones to ten.

Count by ones to fifteen.

Count by ones to twenty.

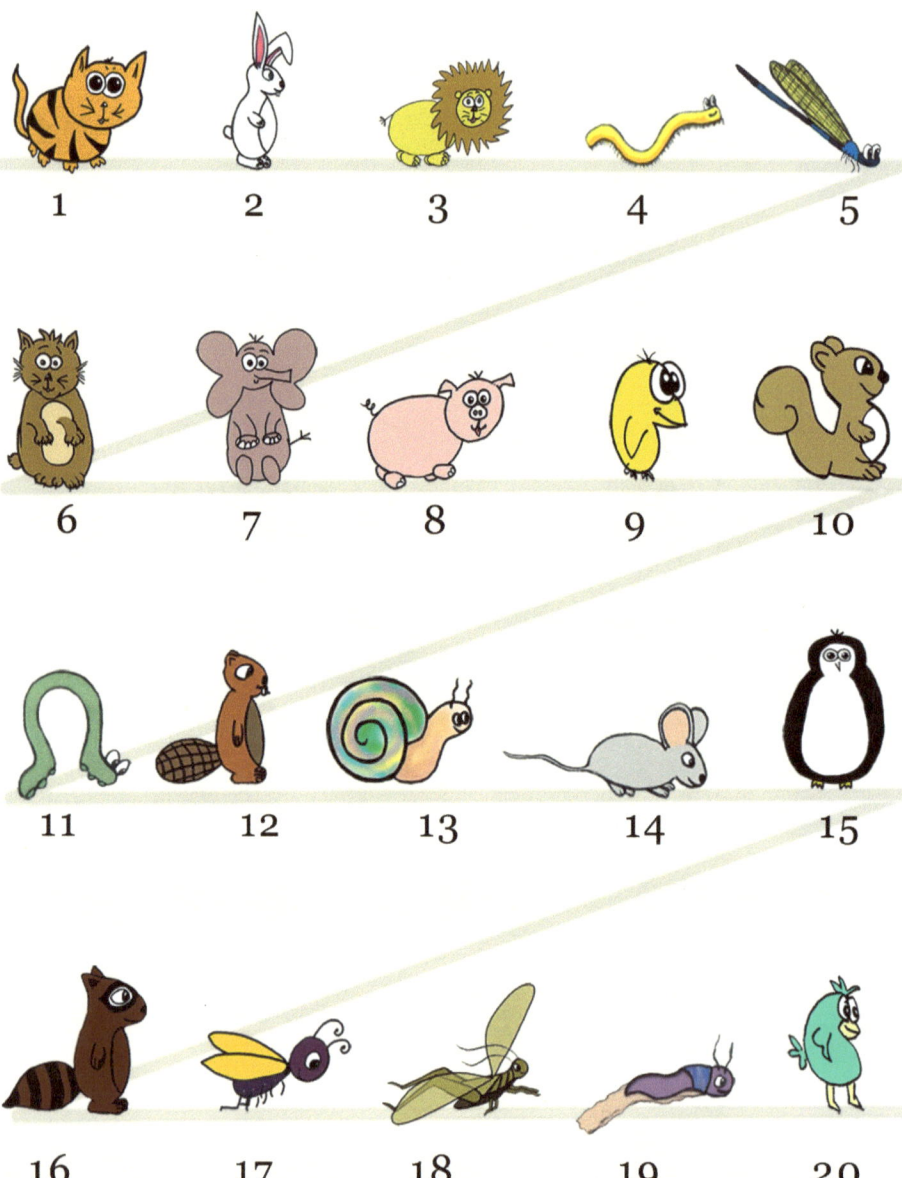

Count by ones to fifteen.

Count by ones to twenty.

**It's also true
you can count by two.**

(We count eyes on froggies.)

2
4
6
8
10
12
14
16
18
20

Count the 👀 in pairs of 2.

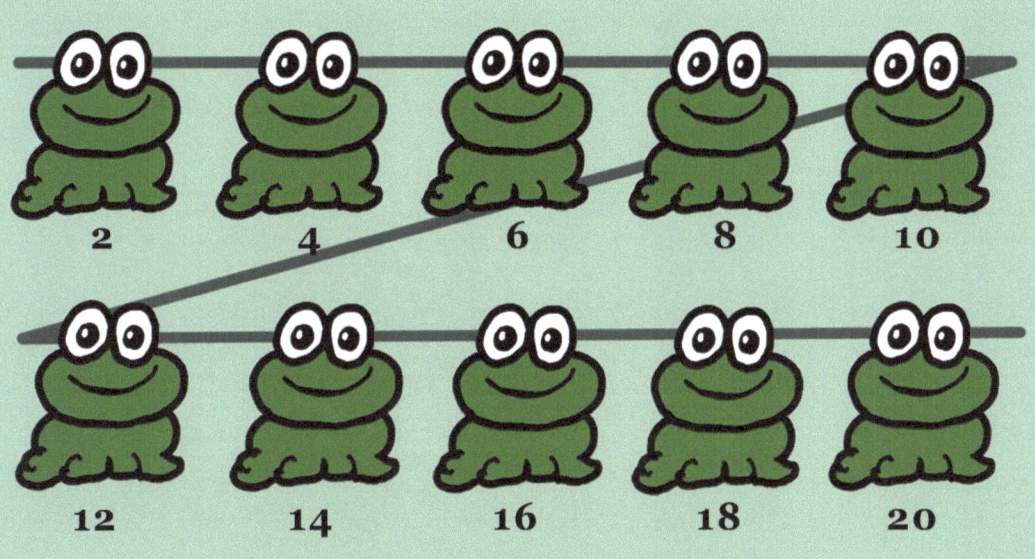

Since it is free,
you can count by three.

(Bees have black stripes.)

3
6
9
12
15
18
21
24
27
30

Count the 〰 in groups of 3.

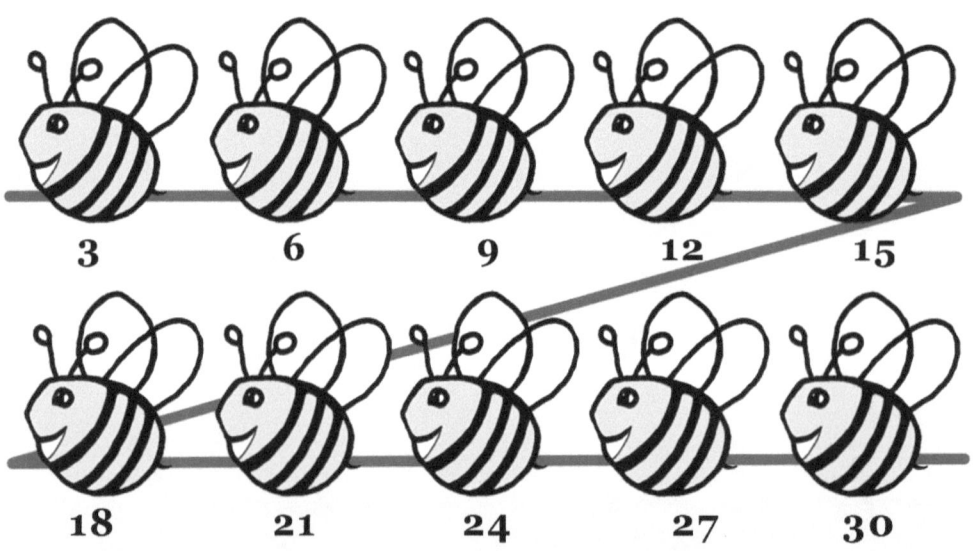

3   6   9   12   15

18   21   24   27   30

It's also true
you can count by two.

(We count eyes on froggies.)

Count the  in pairs of 2.

**Since it is free,
you can count by three.**

(Bees have black stripes.)

Count the  in groups of 3.

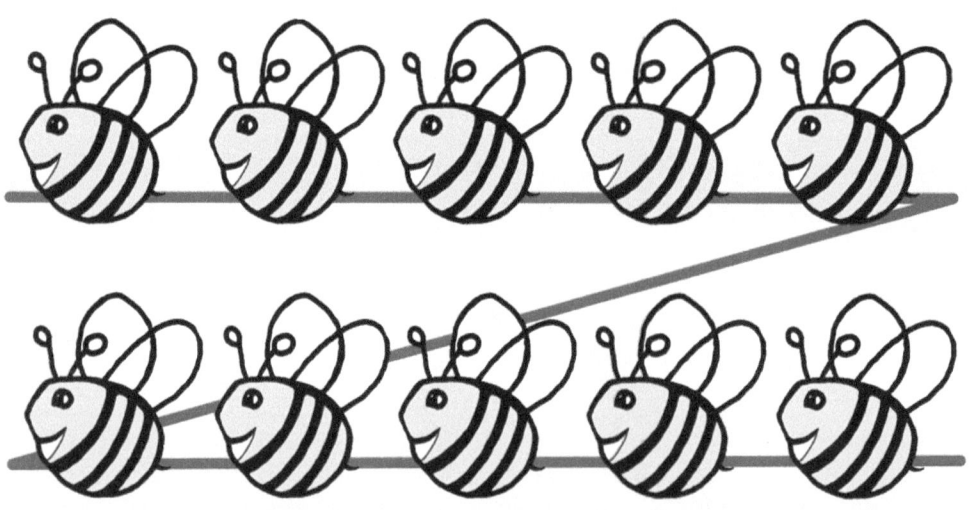

Paws off the floor,
Count them by four.

4
8
12
16
20
24
28
32
36
40

(We count bear paws.)

Count the  in groups of 4.

4　　8　　12　　16　　20

24　　28　　32　　36　　40

Before you squirm,
Count the stripes on a worm.

(Nickles are counted by fives.)

5
10
15
20
25
30
35
40
45
50

Count the 🪱 in groups of 5.

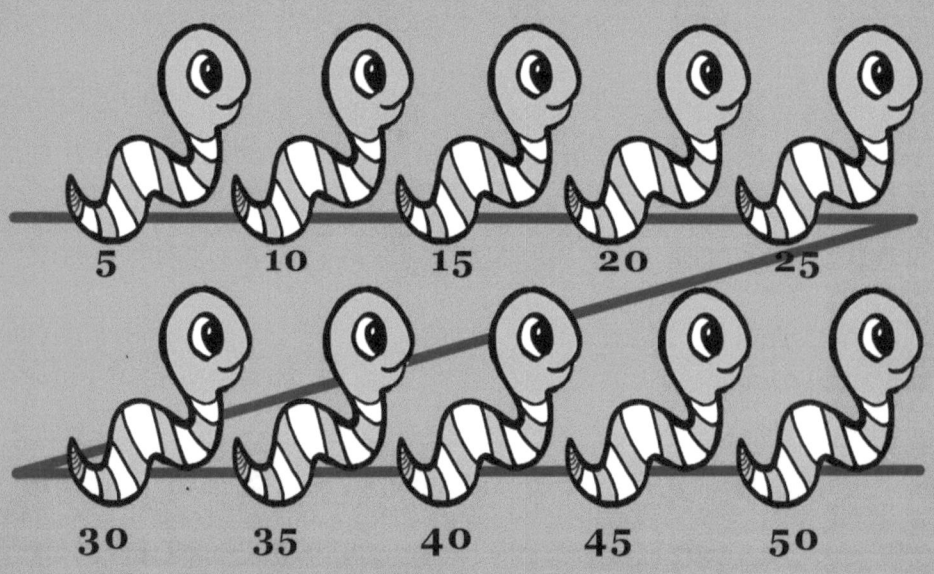

5   10   15   20   25

30   35   40   45   50

Paws off the floor,
Count them by four.

(We count bear paws.)

Count the  in groups of 4.

**Before you squirm,
Count the stripes on a worm.**

(Nickles are counted by fives.)

Count the  in groups of 5

**To get some kicks,
we will count by six.**

Count spots on butterflies.

Count the in groups of 6.

	06
	12
	18
	24
	30
	36
	42
	48
	54
	60

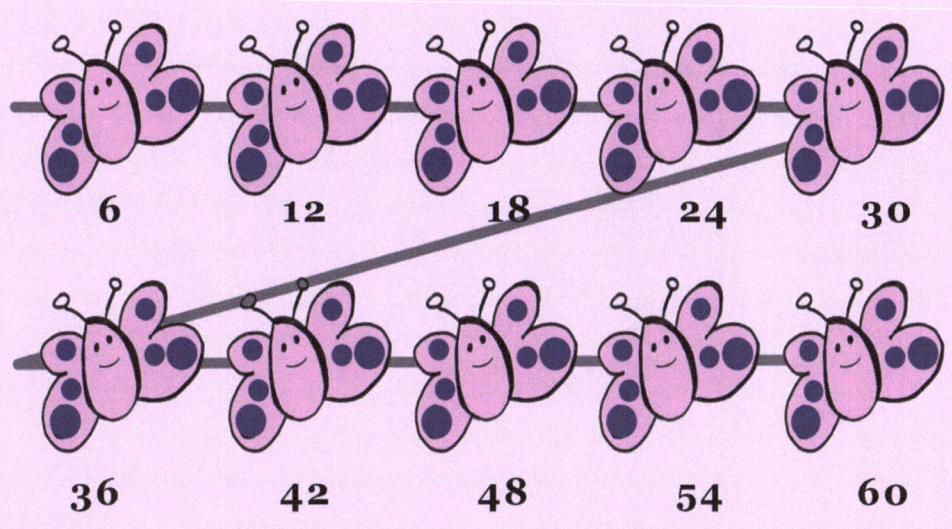

6   12   18   24   30

36   42   48   54   60

**We've now begun
to count by sevens.**

07
14
21
28
35
42
49
56
63
70

Weeks have seven days.

Count the in groups of 7.

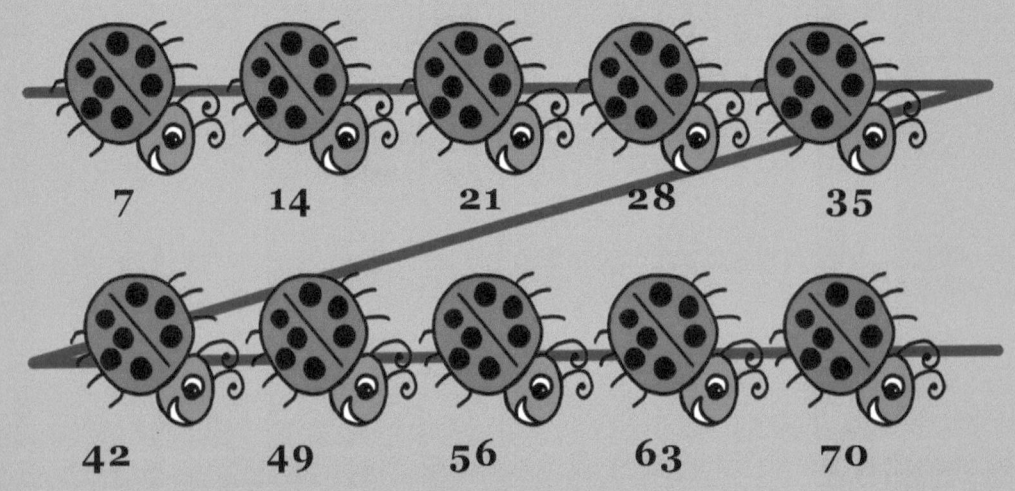

7   14   21   28   35

42   49   56   63   70

To get some kicks,
we will count by six.

Count spots on butterflies.

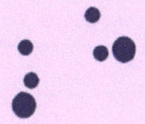

Count the in groups of 6.

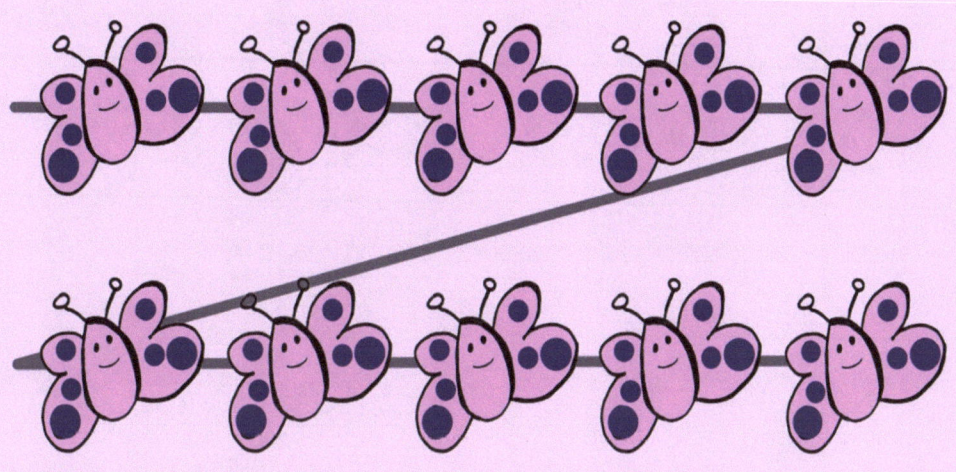

We've now begun
to count by sevens.

Weeks have seven days.

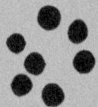

Count the in groups of 7.

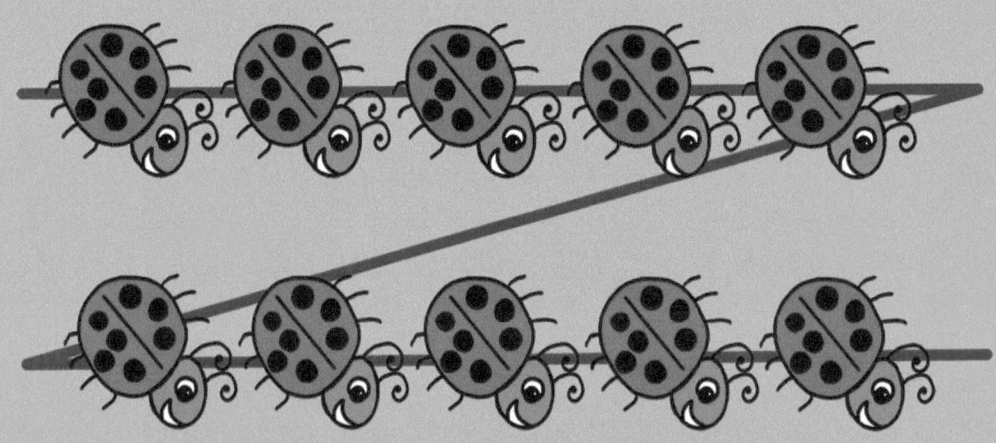

**Some folks can't wait
to count by eight.**

08
16
24
32
40
48
56
64
72
80

Spiders have eight legs.

Count the in groups of 8.

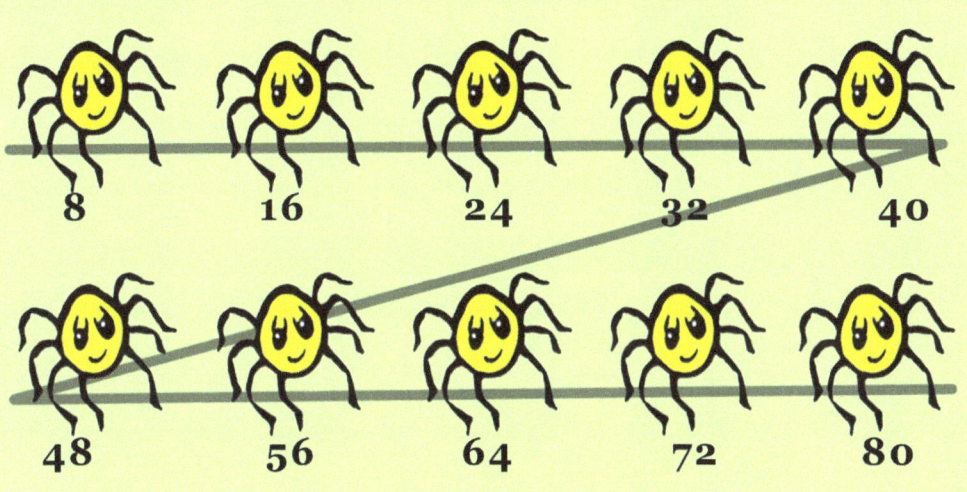

8   16   24   32   40

48   56   64   72   80

**It's easy and fine,
to count by nine.**

**Baseball players are
counted by nines.**

Count beetle dots.

09
18
27
36
45
54
63
72
81
90

Count the 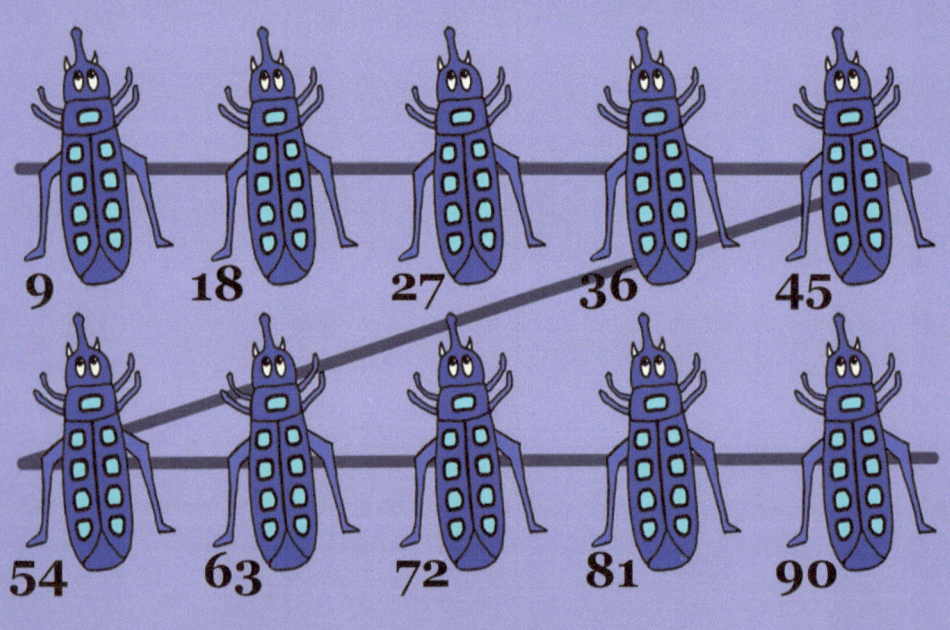 in groups of 9.

9 18 27 36 45

54 63 72 81 90

**Some folks can't wait
to count by eight.**

Spiders have eight legs.

Count the  in groups of 8.

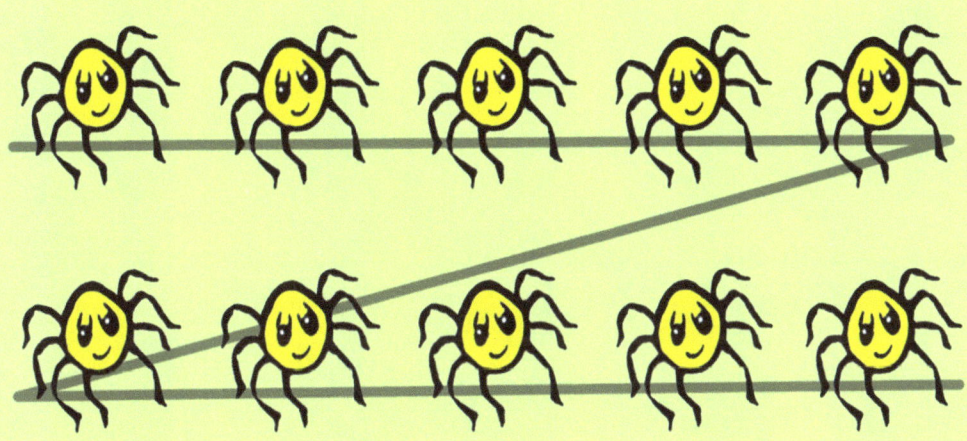

**It's easy and fine,
to count by nine.**

**Baseball players are
counted by nines.**

Count beetle dots.

Count the in groups of 9.

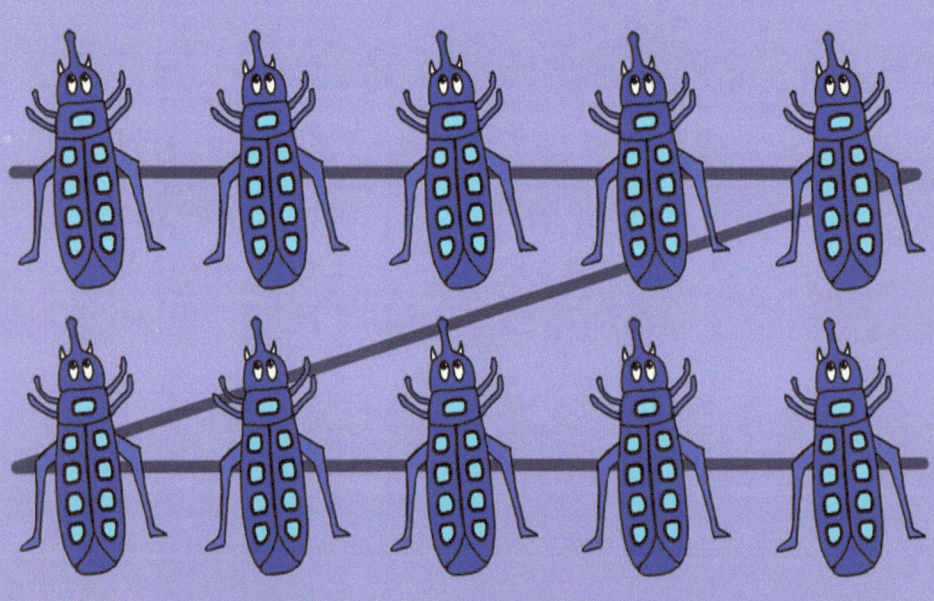

**Now and then
we must count by ten.**

10
20
30
40
50
60
70
80
90
100

We count dimes by tens.

Count the in groups of **10**.

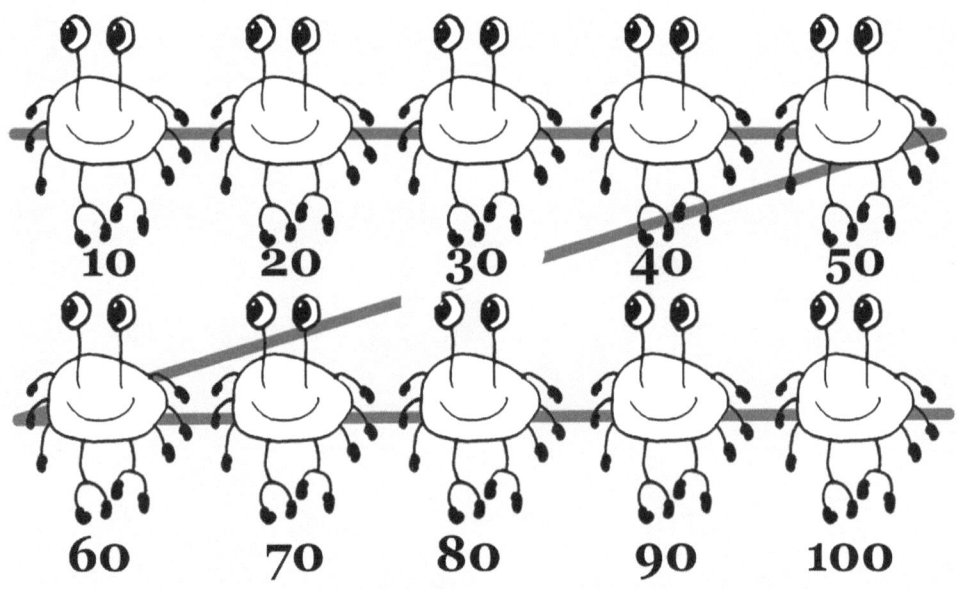

10   20   30   40   50

60   70   80   90   100

It's easier than sevens
to count by elevens.

Football players are
counted by elevens.

11
22
33
44
55
66
77
88
99
110

Count legs in groups of 11.

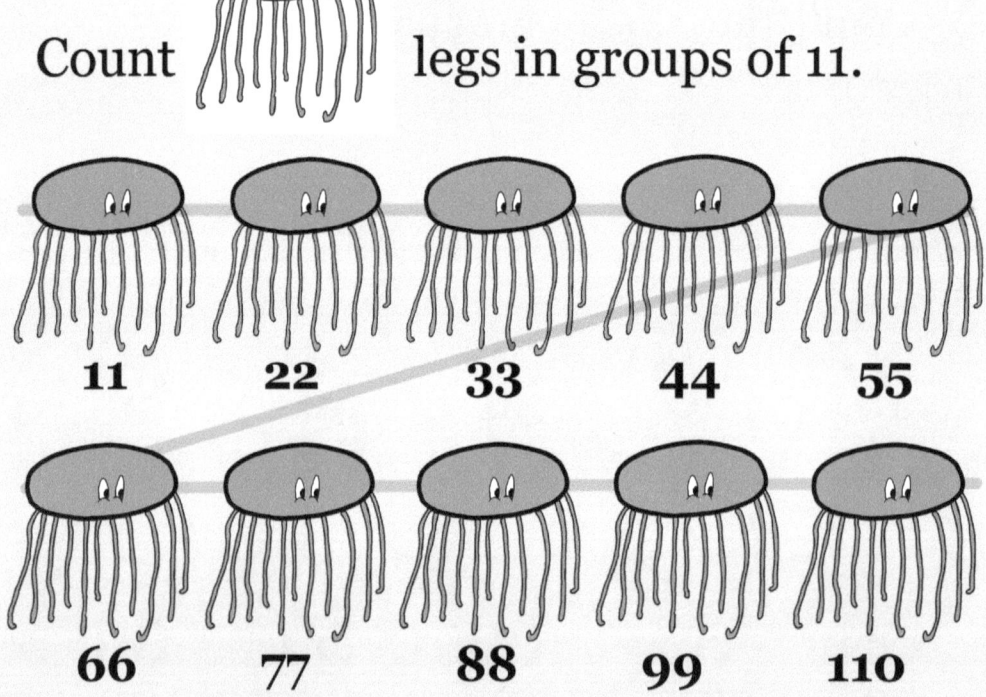

11   22   33   44   55

66   77   88   99   110

Now and then
we must count by ten.

We count dimes by tens.

Count the  in groups of 10.

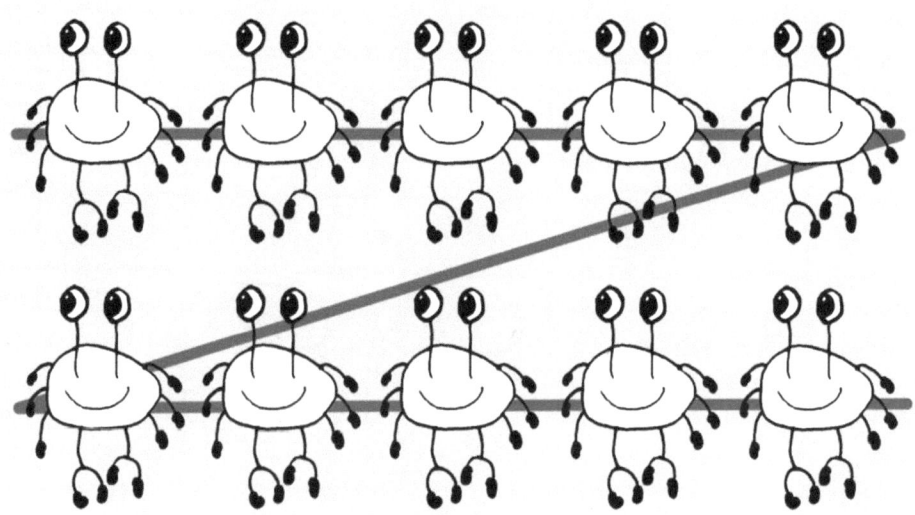

It's easier than sevens
to count by elevens.

Football players are
counted by elevens.

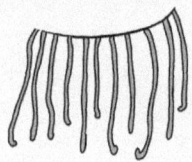

Count 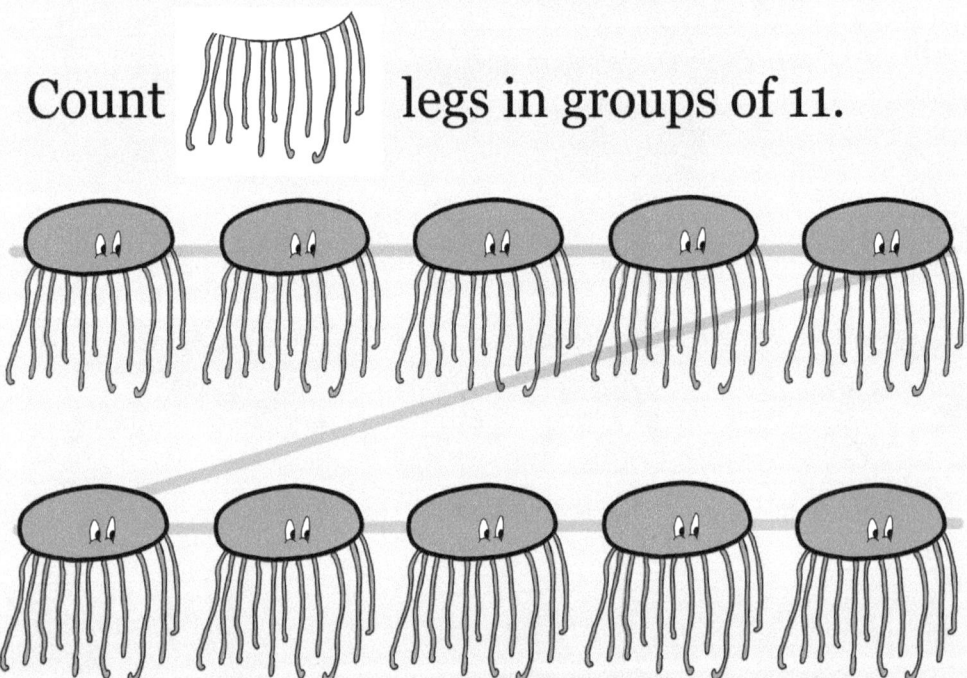 legs in groups of 11.

Twelve is a dozen
eggs in a carton.

Count the legs on the
amoeba.

12
24
36
48
60
72
84
96
108
120

Count  in groups of 12.

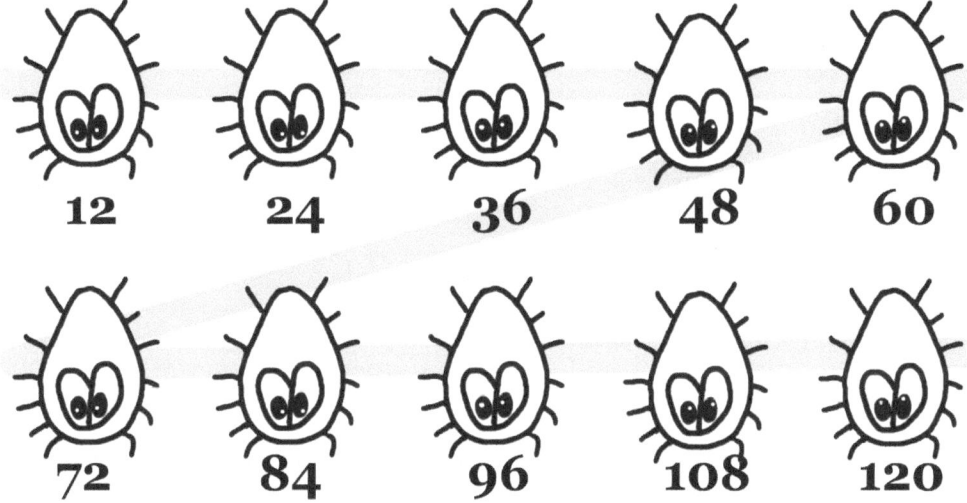

Kids Count by Groups

Encourage your child to look for patterns in the groups. This is especially dominant in the fives and tens, but also in the sixes and eights. The strongest pattern is the tens, with the nines a close second. Point out that the two numbers in the answers for the nines always add up to nine. Also, when multiplying by nine, the answer will always start with the next lower number. 9x7=63 (The 6 is one lower than the 7)

Counting up means you count from smaller to **larger.**

05
10
15
20
25
30
35
40
45
50

Counting down means you count from **larger** to smaller.

50
45
40
35
30
25
20
15
10
05

Twelve is a dozen
eggs in a carton.

Count the legs on the
amoeba.

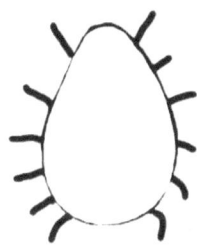

Count  in groups of 12.

READING ANYONE?

We hope later on to have a short reading program similar to this, which a parent can read with his child, to help learn the phonic sounds. A small reader, of phonic and sight words, would go with it.

If you found this little book useful, please give us a review on Amazon. They help spread the word about our products.

Kids Count by Groups

Bedrock Video Productions LLC

Ones		2	3	4	5	06
1	2	2	3	4	5	06
3	4	4	6	8	10	12
5	6	6	9	12	15	18
7	8	8	12	16	20	24
9	10	10	15	20	25	30
11	12	12	18	24	30	36
13	14	14	21	28	35	42
15	16	16	24	32	40	48
17	18	18	27	36	45	54
19	20	20	30	40	50	60

07	08	09	10	11	12
07	08	09	10	11	12
14	16	18	20	22	24
21	24	27	30	33	36
28	32	36	40	44	48
35	40	45	50	55	60
42	48	54	60	66	72
49	56	63	70	77	84
56	64	72	80	88	96
63	72	81	90	99	108
70	80	90	100	110	120

You'll enjoy these Rande titles.

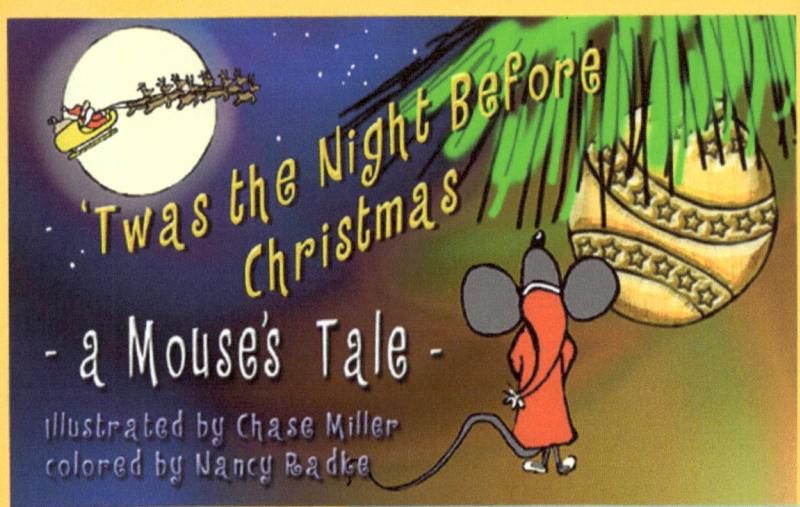

'Twas the Night Before Christmas
- a Mouse's Tale -
Illustrated by Chase Miller
colored by Nancy Radke

In print and on Kindle.

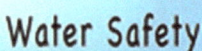

Rande's First Swim
by Chase Miller   Water Safety

Rande's Snow Day
By Chase Miller

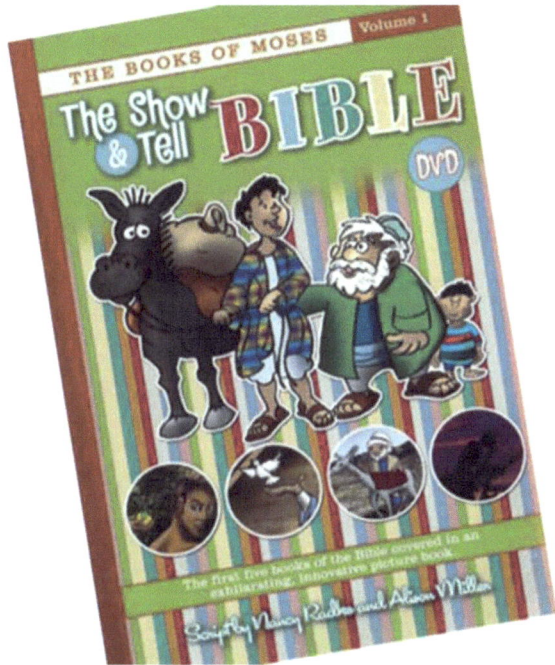

See all 2 hours of the
Books of Moses
Genesis
Exodus
Leviticus
Numbers
Deuteronomy

English/Spanish

Coloring books
&
Activity books

www.showandtellbible.com

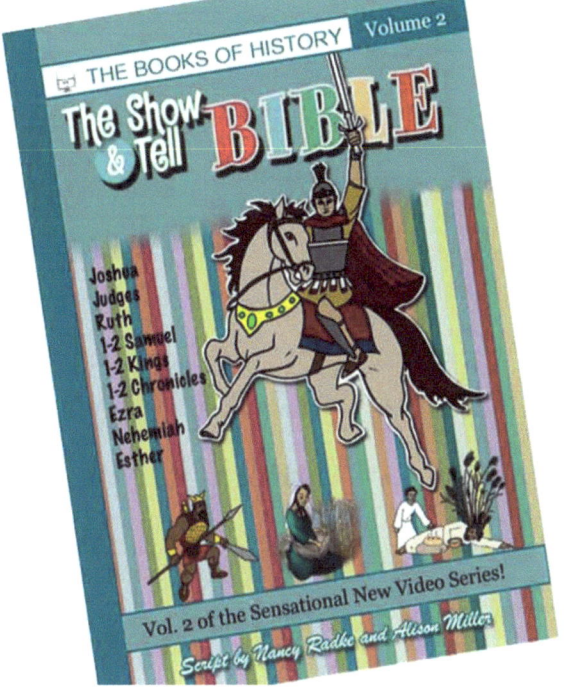

Volume 2
Books of History
Joshua
thru
Esther

(3 hours)

English/Spanish

Coloring books
&
Activity books

www.showandtellbible.com

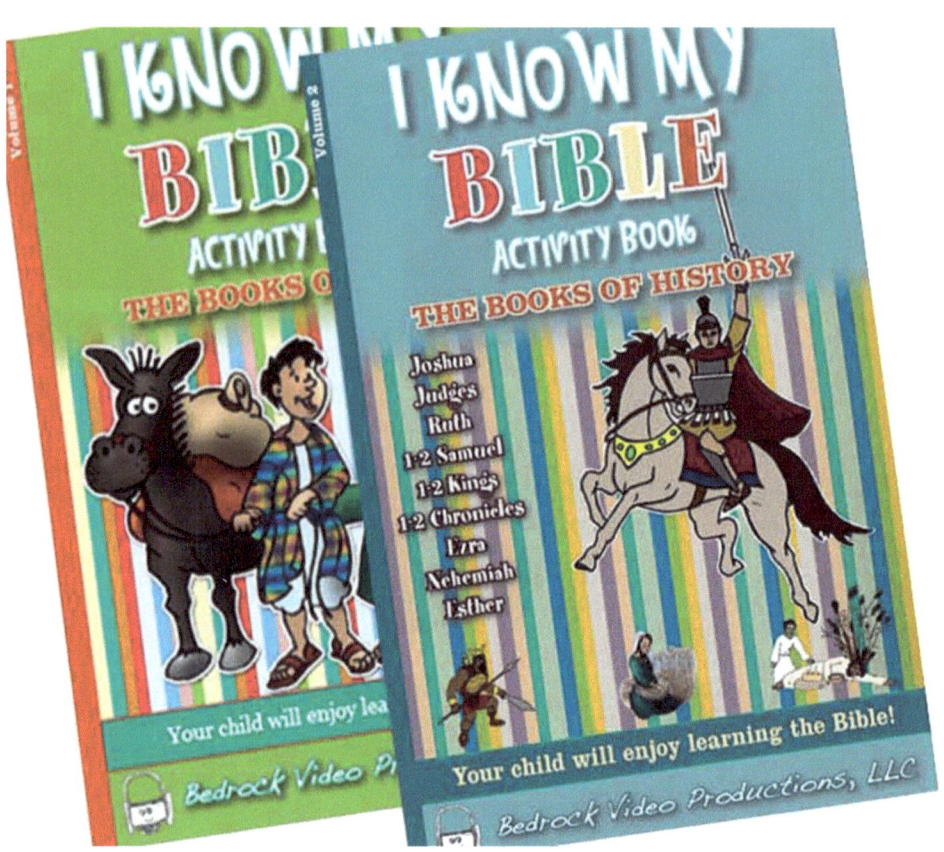

Activity Books
&
Coloring Books

showandtellbible.com

www.ingramcontent.com/pod-product-compliance
Lightning Source LLC
Chambersburg PA
CBHW040753200526
45159CB00025B/2082